Convert your car into a Mini-Camper

Step by step guide for the camper-conversion of your car

Jörg Janßen- Golz

Table of Contents

Introduction

I love camping! For me there is no better holiday than being outside and living by my own set of rules and by my own clock. I just love it when I don't have to follow any breakfast- or dinner times or opening hours for the pool at a hotel. If I don't like the place I'm at, I just gather my stuff and drive to a better destination. Camping is getting more and more popular and every year more people follow this way of independence during their holidays.

After many years of camping in a tent, where I had to endure rain, storm, cold and heat, I was ready for an upgrade. Therefore I searched for a suitable car, which is serving as a normal car during my everyday life and that could turn into a Minicamper for my holidays and short trips. The basic idea behind this project was to sleep safe and comfortable in the car and having enough storage space for all my camping gear at the same time.

My choice fell on a Volkswagen Caddy, which for me is the perfect combination of a spacious commercial vehicle and a practical car for driving around during the day. The Caddy is agile in normal traffic and small enough to fit in every parking space. But turned into a "rolling bed" it offers enough space for two grown people sleeping comfortable side by side.

My goal was to easily convert the Caddy, without making any permanent changes at the car itself so I could switch back and forth between the two sets.

The car on which this conversion is based is a 2012 VW Caddy III (not the Maxi version). The basis for the "bed" is a simple box-system of 4 single boxes with lids. When closed, the boxes form an even footprint of a good 2 meters in length and 1,3 meters in width and therefore enough space to fit two grown-ups.

!!!ATTENTION!!! For this conversion the back seat row has to be removed, so that just the driver and co-driver seats remain.

After having read through a lot of internet forum I spoke to a carpenter and asked if he could make me an offer for building the boxes for me. His initial offer was round about 1.500 €. Since this was way out of my budget, I decided to build the boxes myself. With good planning, patience and good but simple tools I managed to build the system all by myself in my garage and with a budget of under 500 €. I chose high quality material, especially the wood, which was the most expensive part. If you chose to use cheaper wood, this conversion can be done for less than 300 $.

This construction manual will enable you to convert your car into a Minicamper, using simple, everyday tools (which you either have yourself or can borrow from a friend). This guide offers you the complete material and tool lists, a cost overview, exact measures, technical drawings and pictures of the single steps. All materials are available in your local hardware store or can be ordered via the internet.

This guide will give you the chance to turn your dream of a Minicamper into reality. I want to encourage you to give it a try, since I managed to build this conversion all by myself with simple tools in my garage. And the result I got gives me the opportunity to be on the road and sleep safe and comfortable even in bad weather conditions. I've been on several festivals, weekend trips and a two week trip to Spain. I never slept better during a camping trip.

I would like to remind you that the conversion of your car lies in your own hand. I can only speak for myself and my own experience and progress. For me this conversion worked really well and it fits perfectly into my car, without a single scratch on the car itself. I do not take any responsibilities for any damage you caused by using this guide and I will not give any claims of compensation.

!!! Please be aware that I am from Germany, therefore I am only familiar with the metric system!!! All of my measurements are in Centimeters and Millimeters, NOT in Feet or Inch. Please excuse any misspelling; I will try my best to make this guide as simple as possible to understand.

Just a note on the site: Take your time to read through the entire guide at least once before starting the construction. This will give you a complete overview of the whole process and a lot of things will fall into place and seem more logical to you.

You can find a video on how to install this system on **Youtube.**

Just search for "

"Turn your car into a Minicamper"

I wish you as much fun as I had while planning and building this conversion. It is a great experience when in the end you build something with your own hands that is so useful and gives you so much joy in your life!

You can follow my trips with my Minicamper on my blog www.vizslaugh.de

See you on the road!

Joerg

Material list

1. Multiplex panels 18mm thick (Measures for cutting process in chapter 1)

For the material I chose multiplex panels as a conscious decision. Multiplex is very durable and in addition it is water repellent. You could also choose plywood, which is not water repellent, but a lot cheaper, if you are interested in cutting your budget even further.

The 18mm thickness I chose out of a gut feeling. This makes the boxes very robust, but also quite heavy. The boxes are in my eyes almost indestructible and will last for a very long time. The stability will also be given with 15 mm thickness. But be aware that if you pick 15mm you have to pick thinner screws than I did.

Costs (depending on the hardware store): about. 300 $

2. Plywood screws, Torx headed 4,0 x 45 mm
pieces: about. 200
costs: about. 6 $

3. Plywood screws, Torx headed 4,0 x 20 mm
Pieces: about 50
Costs: about 4$

4. Hexagon screw 8 x 55 mm
Each with 2 washers and one wing nut
(To connect the 4 boxes with each other)
Pieces: 6
Costs: about. 3 $

5. Hexagon screw, 4 x 45 mm
Each with 1 washer and one wing nut
(To connect the metal brackets with the boxes 1&2)
Pieces: 4
Costs: about. 2 $

6. Hexagon screw, 5 x 35 mm
Each with 1 washer and one nut
(To connect the pipe clamps to box 3&4)
Pieces: 8
Costs: about. 3 $

7. Carriage bolt (mushroom head bolt) 8 x 55 mm
Each with 1 washer and one nut
(To connect the headboards to box 3&4)
Pieces: 4
Costs: about 3 $

8. Carriage bolt (mushroom head bolt), 4 x 35 mm
Each with 1 washer and one nut
(To connect the Folding Brackets with boxes 3 & 4)
Pieces: 4
Costs: about 2 $

9: Pipe Clamps Ø25 mm
Pieces: 4
Costs: about 6 $

10. Surface Mount Hinge with spring (Length 103mm width 44mm)
Pieces: 8
Costs: 1,50 $ each

10. Folding Brackets 341 x 136 mm
Pieces: 4
Costs: about 30 $

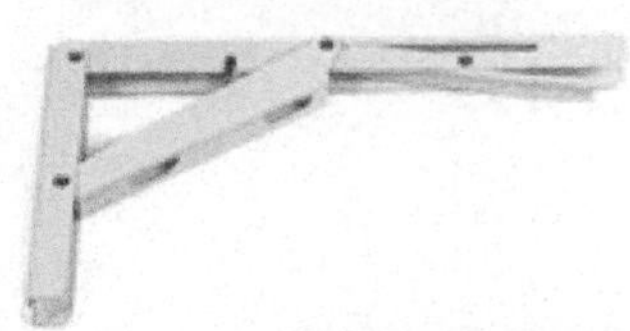

11. Metal Elbow Brackets 105 x 105 mm
Pieces: 2
Costs: about 3 $

Tools list

Drilling machine

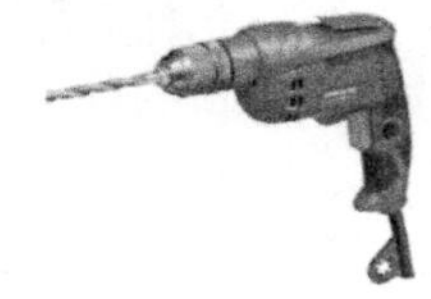

Jigsaw and new sawblades

Wood Drill 8mm, 4mm, 5mm and 3mm

Cordless Screwdriver

Torx- Bit

Countersink

Set Square

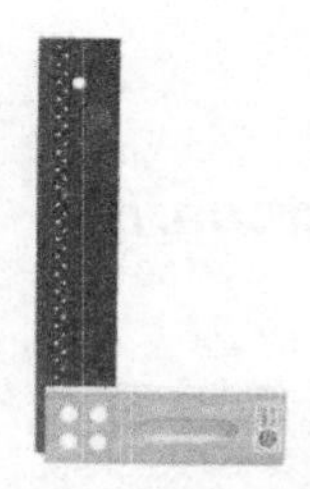

Hole saw 45 mm diameter

Delta Sander or Sanding paper and sanding block

Spanners

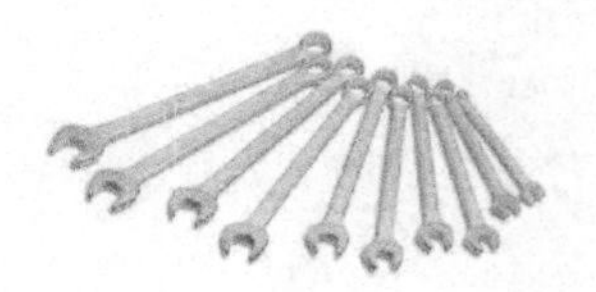

The end- result, all 4 boxes together with headboards

1. Cutting of the wood

The usable space of the trunk is 112cm in width. When the front seats are pushed all the way to the front and the backrest fully inclined, the usable length is 194 cm. Therefore the footprint of the box system will be 112cm x 194 cm.

This space is divided as follows:

Boxes 1 to 4, Center pieces 1 and 2 and headboards 1 and 2 (see Illustration 1).

Feel free to choose the height of the boxes as you like. I have chosen the height of about 50 cm on purpose. For once, this height is needed to store my cooling box in one of the boxes. In addition, when choosing this height, the mattresses will be just high enough that you will lie between the side windows. This will give you a few more centimeters of space on each side and therefore you will sleep more comfortable. This results in a little bit less headroom to the roof, but it is still just as comfortable. My advice is to really think about what you have to store into the boxes and choose the height according to your needs.

In the beginning I measured everything that would later be stored in the boxes. My folding chairs and my folding table have a length of about 90 centimeters, so I calculated the boxes 1 and 2 with a total length of 100 centimeter, so that everything fits right in.

The space between box 1 and 2 was chosen to be 44 cm in width, so that a folding box or an aluminum box for extra storage fits right in.

Therefore the outside measures of the boxes are as follows (see illustration 1):

Box 1: 28 x 100 x 47 cm (width x length x height)

Box 2: 40 x 100 x 47 cm (width x length x height)

Box 3 & 4: 55,5 x 47 x 47cm (width x length x height)

Headboards 1 & 2: 55,5 x 47 cm (width x length)

Center Panels 1 & 2: 45,8 x 50 cm (width x length)

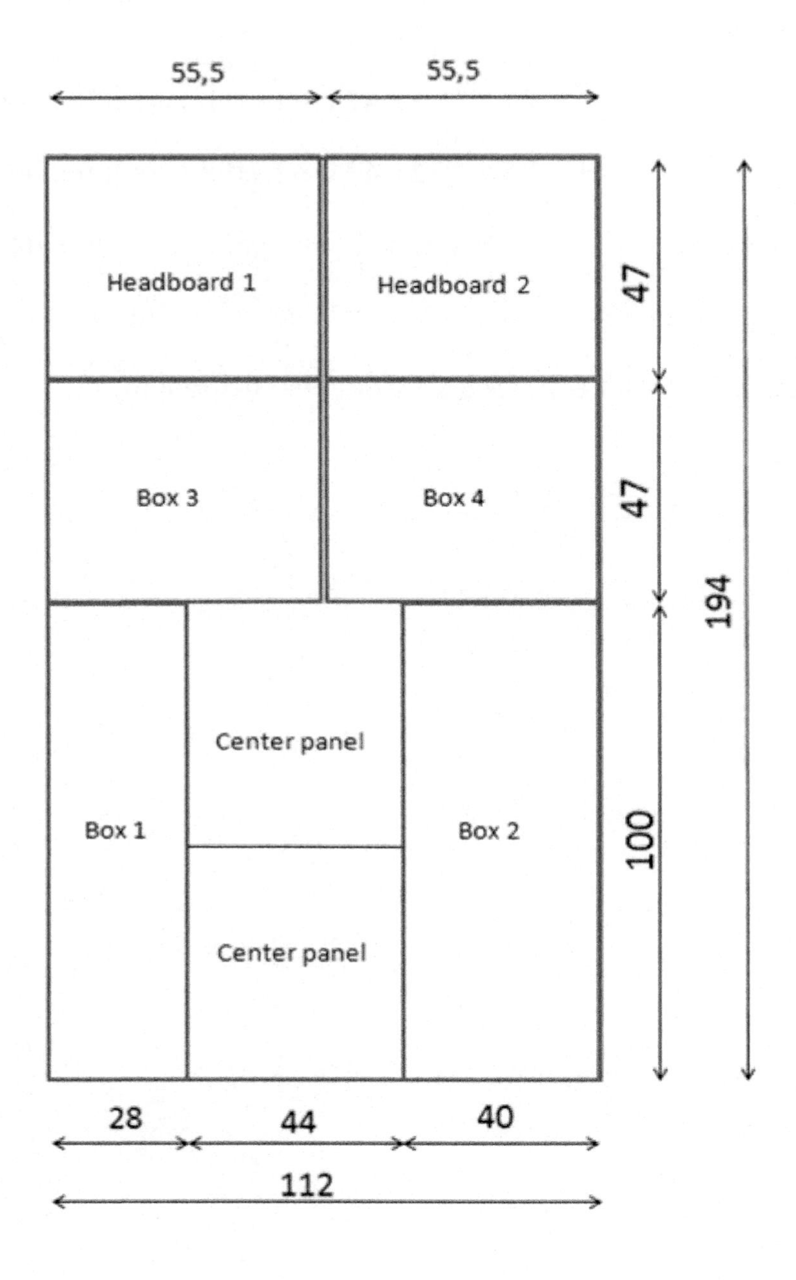

Illustration 1

Since each box has its own lid, this concludes to the following material list for the multiplex panels:

Pieces	Length (in cm)	Width (in cm)	Use for
1	100	28	Floor panel box 1
1	100	27,1	Lid box 1
4	100	43,4	Side panels box 1&2
2	100	40	Floor panel box 2
1	100	39,1	Lid box 2
2	24,4	43,4	Front&Back panel box 1
2	36,4	43,4	Front&Back panel box 2
6	55,5	47	Floor panel + lids box 3&4+ headboards
4	47	43,4	Side panels box 3&4
4	51,9	43,4	Front&Back panel box 3&4
2	50	45,8	Center panels 1&2

You can take this list to your local carpenter or your favorite hardware store and ask for an exact cutting.

!!! This part is crucial, be aware that the measurements really have to be exact, otherwise the boxes will not fit in the end!!!

While you are at the carpenter, ask him to cut a so called chamfer (Illustration 2) on the outside (!!!!!) of the lids for box 1 and 2. This will prevent the lids from scratching against the interior of your car when you open them. Also put a chamfer on the lid of box 3 and 4 on the side where they touch box 1 and 2. This will allow you to open the lids much smoother than without the chamfer.

Just take the picture of the chamfer to the person who cuts the wood. He will know how this chamfer has to be cutted.

Illustration 2

Tongue on the lids and Center Panels

You will need a so called tongue on the lids for the boxes 1 and 2 and on the sides of the Center Panels. This will ensure a good fit and additional stabilization between the two parts (see Illustration 3 & 4).

The lids will be tongued on the upper side; the Center Panels will be tongued on the bottom side. The tongue is 9x9 mm wide, exactly the half of the panel thickness.

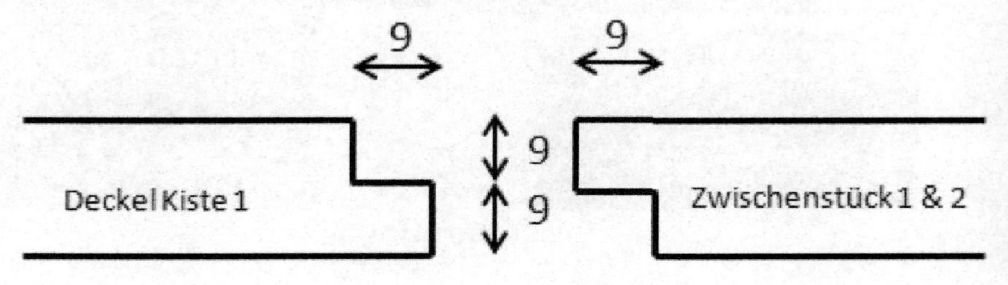

When all parts are fit together, it looks like this:

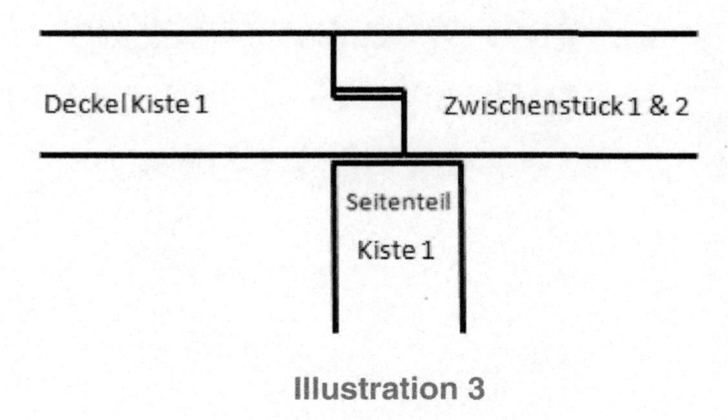

Illustration 3

Illustration 4

2. Material inspection and assignment

The first thing you should do immediately (!!) after the cutting is to check the correct measurements of the panels. You should do this, while you are still at the carpenter or hardware store. This is the best chance to ensure, that all box can be assembled correct. If the measurements are incorrect or slightly inaccurate, this is the moment to speak to the person who did the cutting and ask for a correction.

When everything is correct, it is time to assign each panel to the right box. Write onto the cut surface of each panel the number of the box they belong to. This way you can always see immediately which panel and which box you are currently working on. Do the same thing for the headboards and the Center Panels.

To keep a better overview, put together the panels of each box and mark them with a so called “carpenter’s triangle” (see Illustration 5). This makes it clearer which panel you are currently working on and reduces the risk that you mix up the different parts. By using this method you can also see if you holding a side panel or a front or back panel in your hand.

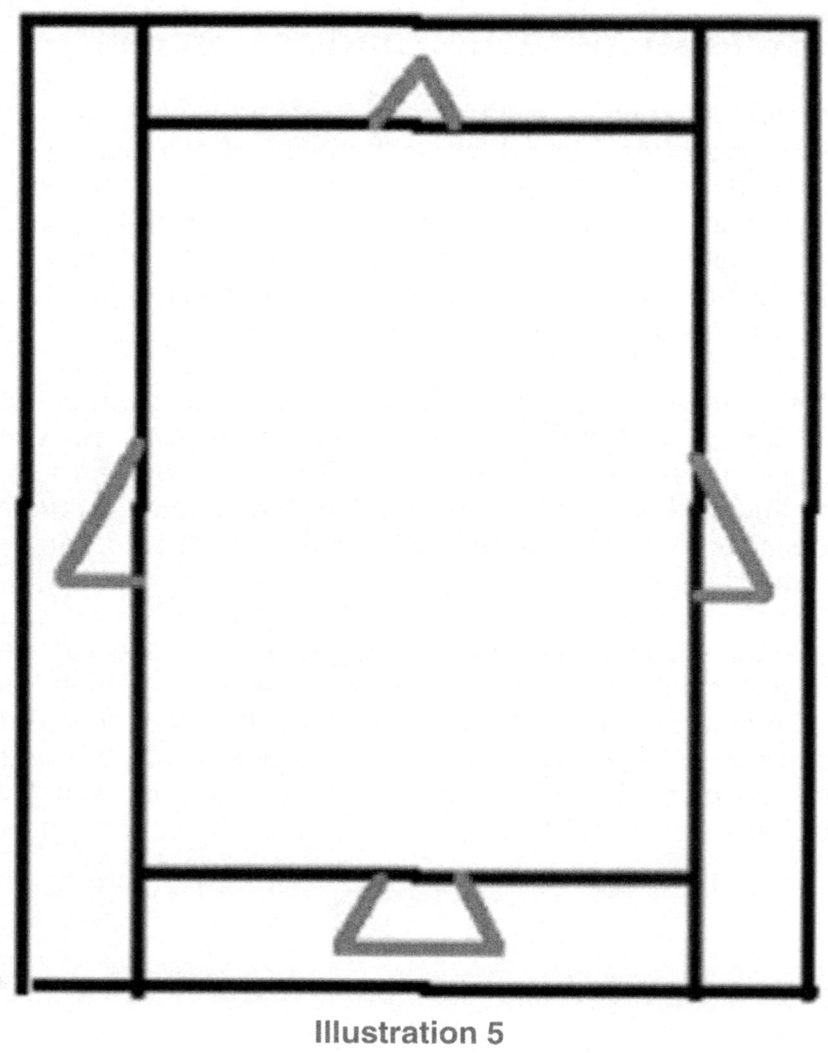

Illustration 5

3. Pre- drilling of the floor panel

Start your project with the floor panels of the boxes 1 - 4. All the side panels are getting screwed to the floor panel from below later on. To make this a lot easier, it is important to pre-drill the holes for the screws. This will also prevent the panels to snap or break.

For this you should mark the drilling holes on the bottom side of the floor panel. Use a distance of 9mm from the outside of the panel (see Illustration 6). That way you can be sure, that the screws take hold right in the middle of the side panels.

Make sure, that the smooth surface of the multiplex panel points outside later on, since this is the waterproof side.

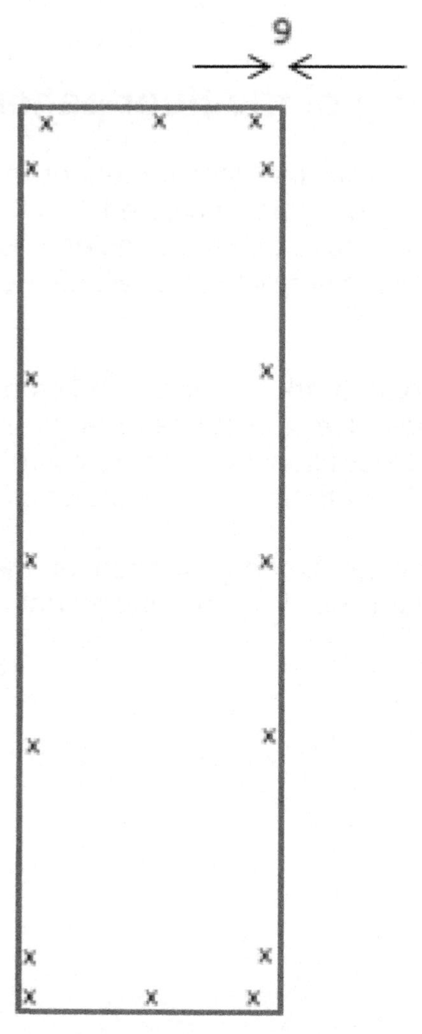

Illustration 6
Floor panel box 1

Mark the drilling holes all around the panel. Chose the distance between each hole, approx. 15 cm apart. It is better to use one or two screws more rather than too less. If a pencil does not work for you, you can always use a nail or another sharp tool.

! ATTENTION! Do not drill the holes exactly in the corners! These screws might later get in the way of the screws for the side panels. It is better to keep a few centimeters distance from the edges.

Start drilling the marked holes with the 4 mm wood drill for all 4 floor panels. You should place an old piece of wood beneath the panel to prevent the drill holes from tearing out when breaking through. To prevent the panels from slipping away, you can use a screw clamp to secure the panel to your workbench.

After drilling all the holes you can use the counter sink (see Illustration 7) to sink all the drill holes. This way the heads of the screws will later finish even with the panel and to not stand out.

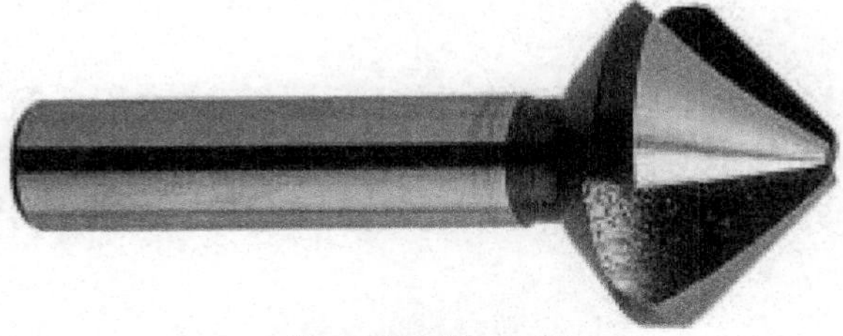

Illustration 7

!! Be aware that at this point the floor panels are not screwed together with the side panels yet. There are several working steps for the panels to be done before!!

4. Pre- drilling of the side panels

Keep working on the boxes 1 - 4. Place all the side, front side and back side panels the way they will be screwed together. Always be aware that the smooth side of the panels is facing outside. In between the steps it is good to check the outside measures of the boxes from time to time. Better measure once too often than once too less. This will give you the secure feeling that you are using the right panels.

Now mark the drilling holes on the outside with a distance of 9mm from the edge (see Illustration 8). Drill the holes with the 4mm wood drill. You should place an old piece of wood beneath the panel to prevent the drill holes from tearing out when breaking through. To prevent the panels from slipping away, you can use a screw clamp to secure the panel to your workbench.

After drilling all the holes you can use the counter sink to sink all the drill holes so deep, that the heads of the screws will later finish even with the panel and to not stand out.

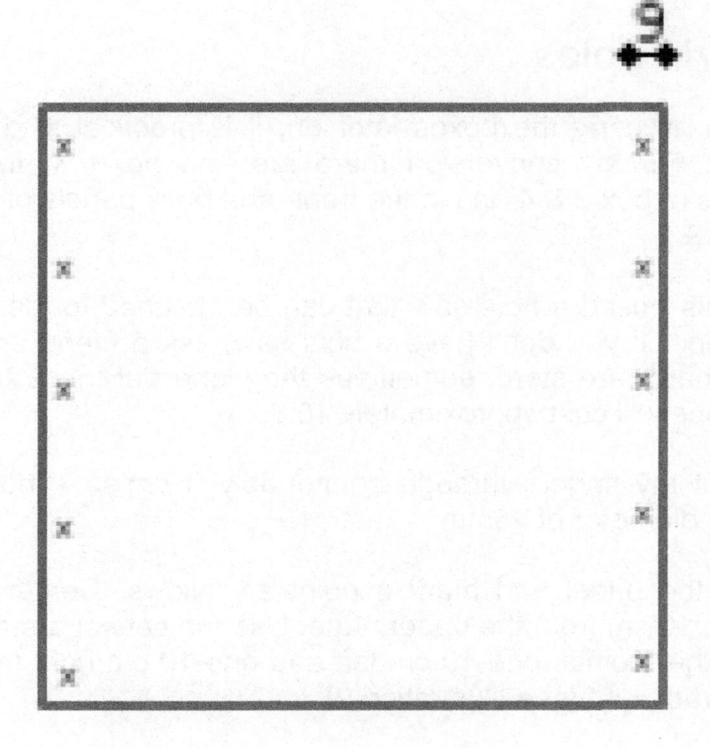

Illustration 8
Side panels box 3 & 4

5. Grip holes

To easily carry the boxes later on, it is practical to drill grip holes. For my conversion there are grip holes in the side panels of box 3 & 4 and in the front- and back panels of box 1 & 2.

For this I used a hole saw that can be attached to the drilling machine. If you don't have a hole saw, ask a friend or at the local hardware store, sometimes they lend out these tools. A new one will cost approximately 10 $.

To put my fingers through comfortably, I chose a hole saw with a diameter of 45mm.

Take the panel and mark a point as follows: Center of the panel, 10 cm from the upper edge. Use the set square to mark two other points, one 10 cm left and one 10 cm right from the centered point (see Illustration 9).

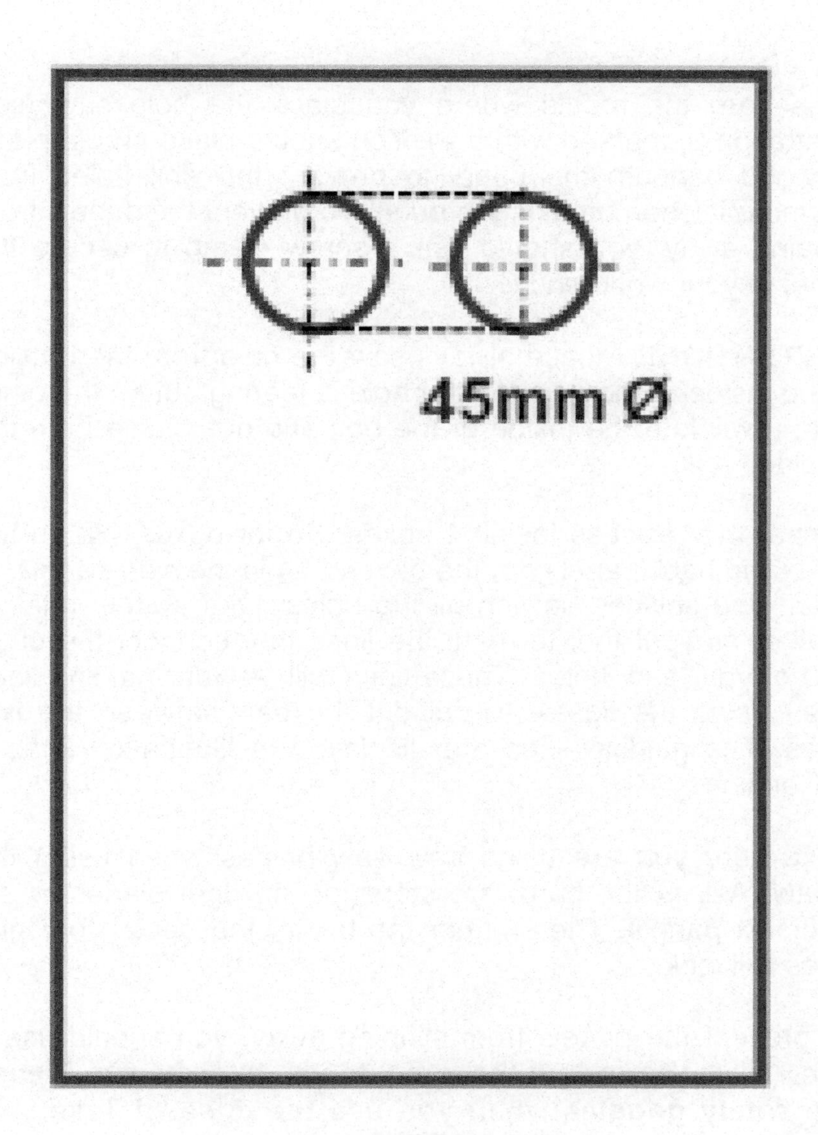

Illustration 9

These are the marks where you place the hole saw (see Illustration 9, marked with a +). You should place an old piece of wood beneath the panel to prevent the drill holes from tearing out when breaking through. To prevent the panels from slipping away, you should use a screw clamp to secure the panel to your workbench.

!!ATTENTION!! Please always saw the hole from the outside to the inside of the panel. If the hole is tearing out on the other side, it will later be inside of the box and not visible from the outside.

Repeat this process for all 4 boxes. So for boxes 1&2 in the front- and back panels, for the boxes 3&4 in the side panels.
When you finished sawing all the holes, take a steel ruler or another straight tool to mark the lines that connect the outer side of your saw holes. These lines will be your guiding lines when using the jigsaw to cut out the part between the two holes. The guiding lines are displayed in Illustration 9 as a broken line.

Make sure you are using new saw blades when using the jigsaw. Ask at the hardware store for the right blades to cut multiplex panels. The sharper the blade, the better your grip holes will look.

To prevent the panels from slipping away, you should use a screw clamp to secure the panel to your workbench. **Please use safety goggles when you use the jigsaw!!** Take your time while using the jigsaw. The more precise you work, the less time you will have to spend on the sanding process afterwards.

Hi, and thank you for still reading my book. If you like it please do me a favor and leave a review for this book on Amazon.

It took me time and effort to write it and you would help me a lot on improving this book for future readers and of course to sell it.

Your review will also help others to decide if they should get the book.

Here is the link to your own review page:

http://www.amazon.de/feedback

Please take two minutes of your time, it would mean a lot to me.

Thank you,

Joerg

6. Air holes

While you are at the drilling and sawing, you should decide if you want to cut air holes into the lids and the headboards. I chose to do this for two reasons.

The first is to save weight. Cutting out the extra material will reduce the weight of the hole system and will end up in fuel savings and less effort while carrying the boxes.

The second reason is to have some air circulation beneath the mattresses. This will reduce the dampness in the mattresses and can prevent mold.

If you decide to go for air holes, the same process as for the grip holes can be used. Take all the lids and the headboards and mark the points where you place the hole saw. Between the air holes you should choose at least 15cm of space to still have enough stiffness. (See Illustration 10, 11+12). That will prevent your panels from breaking, when you are later sleeping on the boxes.

!! ATTENTION!! Be aware to choose the distance between air hole and the edge of the lid wisely. There has to be enough space to mount the folding brackets later on. I chose a distance of about 6 cm.

When you finished sawing all the holes, take a steel ruler or another straight tool to mark the lines that connect the outer side of your saw holes. These lines will be your guiding lines when using the jigsaw to cut out the part between the two holes. The guiding lines are displayed in Illustration 10 as a broken line.

For the sawing process follow the instructions in chapter 5.

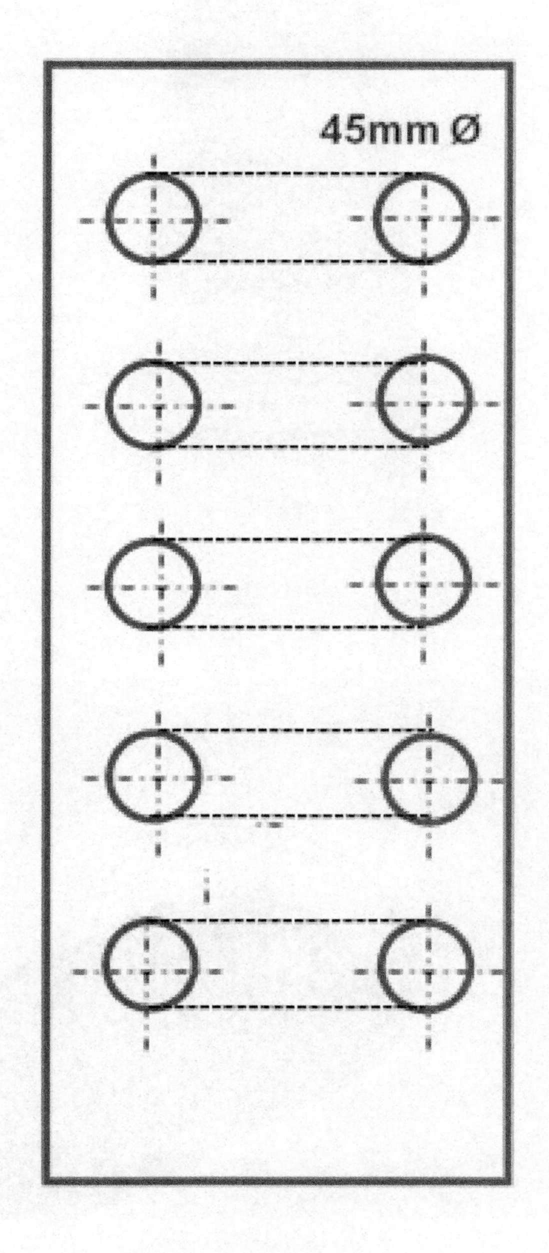

Illustration 10

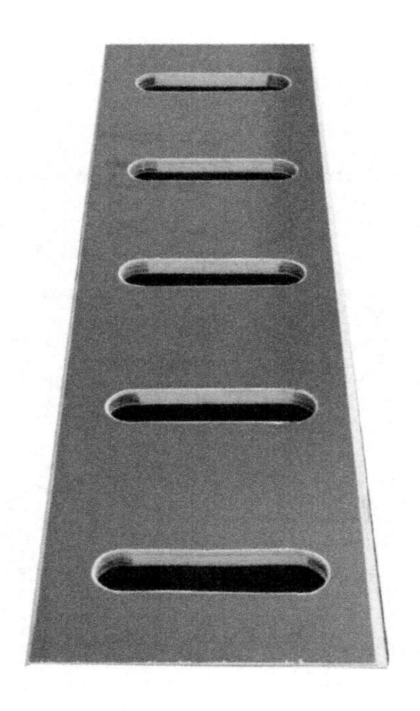

Illustration 11

Illustration 12

7. Sanding

Before you screw together all the boxes, you have to sand all the edges that you will later touch. This means the outer edges, the grip holes, the air holes and the inner edges of the boxes. Sand the edges until they are nice and smooth, so that you won't get any splinters when you later carry the boxes.

If possible, use a delta sanding machine (see tool list). This allows you to reach all the tight corners and into the grip and air holes. This narrowed down my sanding process to 50% of the time.

If you don't have a sanding machine, just take good old plain sanding paper and a sanding block. I used a paper with the grind of 100, which is not too rough and not too smooth. For your safety you should wear a mouth cover during the sanding process.

8. Assembly of the boxes

It can be very helpful to assemble the boxes with two people. I managed it by myself, but it is definitely easier if you have another person to help you. Be sure to use a flat and even surface for the assembly, so that the individual parts can be assembled at a right angle.

Start with the 4 upright panels of the box, so the side panels (the floor panels are assembled afterwards). So take a side panel and a front panel and place them in a 90 degree angle. Make sure that they are placed even, so that no side stands out. The holes that you pre- drilled in step 4 are now the ideal guidelines to screw the two parts together.

Use the Plywood screws, Torx headed 4,0 x 45mm

With a good cordless screwdriver (see tool list) this should be no problem. If the screwing is too hard, you can pre-drill the holes with the 3mm wood drill. Screw together all the side panels, so that you have 4 boxes without a floor panel in the end.

Now it is time to assemble the floor panel. Be aware, that the smooth surface of the floor panel is facing outside.

Use the Plywood screws, Torx headed 4,0 x 45mm

Make sure that all side panels are standing in a right angle. With a little pressure you can push everything into place, so that the boxes are rectangular.

Now you should have 4 boxes in front of you, two quadratic (see Illustration 13) and two rectangular.

Illustration 13

9. Mounting the lids

The lids of the boxes 1 -4 will be attached with so called folding brackets (see Material list). For each lid I used 2 brackets, which works perfectly fine.

For the boxes 1 & 2 the folding brackets will be screwed to that side panel, which is standing on the sidewall of your car. For the boxes 3 & 4 the lids will be screwed to the back panel, which is facing to the rear side of your vehicle.

I assembled the lids of boxes 3 & 4 that way because I wanted to be able to reach inside the boxes from the passenger's seat. This way I can reach my refrigerator box even when I am driving.

When assembling the folding brackets you have to make sure that the distance between the lid and the side panel is correct (see Illustration 14). You should stick to the 2mm distance, so the lid won't get jammed when open it.

Side panel **Lid**

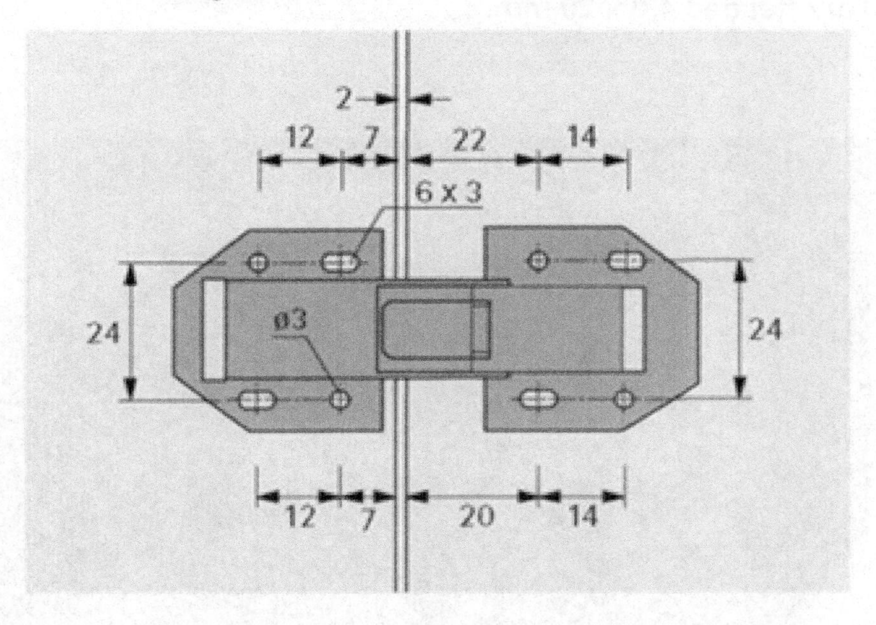

Illustration 14

For the boxes 1 & 2 I chose a distance from the outside of the lid to the folding bracket von about 10 cm (see Illustration 15). Please ensure that box and lid are flush- mounted, so that on the edges, nothing stands out.

For screwing the folding brackets, use the **Plywood screws, Torx headed 4,0 x 20 mm.**

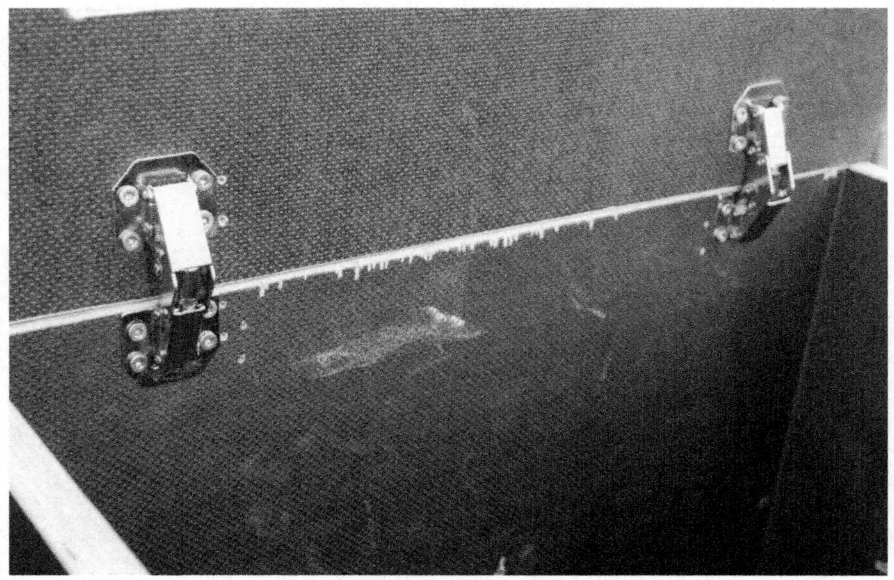

Illustration 15

10. Mounting the headboards

When you've reached your destination with your car and you want to build your bed, the headboards are mounted onto the boxes 3 & 4. To ensure a stable strut for the headboards, I chose 2 space saving and very robust **folding brackets** (see Material list). They can be folded away easily while driving and don't take much space. Later on, you just fold them up and have a steady base for the headboards.

The folding brackets are assembled on the upper side of the front panel on box 3 & 4 (see Illustration 16).

Illustration 16

To provide enough stability I mounted the folding brackets with a **Carriage bolt (mushroom head bolt) 4 x 35 mm, each with 1 washer and one nut.** The hole for this has to be drilled through with the 4mm wood drill. You can see this in Illustration 16, it is the upper of the three screws.

For the other fix points of the bracket you can use the
Plywood screws, Torx headed 4,0 x 20 mm.

!!ATTENTION!!
If you have a center armrest in your car, you have to consider this before placing the folded brackets. Otherwise you will have difficulties opening them when the boxes are installed. If the boxes 3 & 4 are ready, it is best to put them in your car and you can measure exactly where the center armrest starts and where the folding brackets should be placed.

If you have air holes in your headboards, you have to consider this when placing the folded brackets. Otherwise it might be that an air hole is directly there where the headboard lies on the folding bracket later on. So before you mount the folding bracket, hold the headboards where they will be and you will see directly where the bracket has to be placed.

To prevent the headboards from sliding to the sides, I fixed them in place with two **Carriage bolts (mushroom head bolt) 8 x 55 mm and two pipe clamps.**

First, mount the pipe clamps beside the folding brackets, directly on the upper side of the front panel (see Illustration 17).

Use the Hexagon screw, 5 x 35 mm, each with 1 washer and one nut. For this you have to drill the holes with a 5mm wood drill.

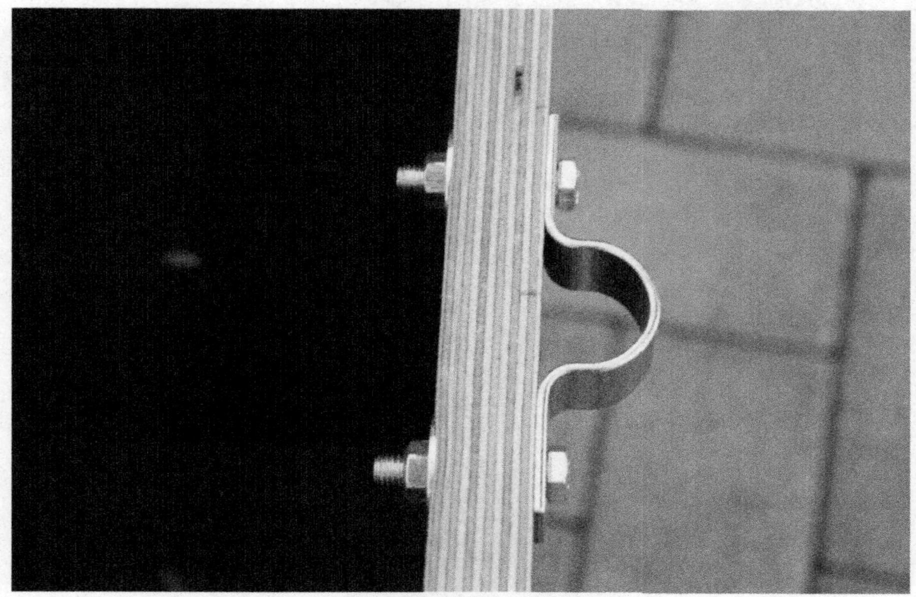

Illustration 17

Now take the headboards and mark the two holes for the two **Carriage bolts (mushroom head bolt) 8 x55mm.** For this you will need the 8 mm wood drill. Later the carriage bolts are just put through the headboards and through the pipe clamps and are fixed from beneath with the washer and the nut (see Illustrations 18+19+20).

Illustration 18

Illustration 19

Illustration 20

11. Connecting the boxes

To prevent the boxes from slipping and sliding in the vehicle, the boxes should be connected. I chose **Hexagon screws 8 x 55 mm, each with 2 washers and one wing nut.**

I decided to use wing nuts so that I don't need any additional tools to connect and disconnect the boxes (see Illustration 21).

Illustration 21

For this, place the boxes on a flat and even surface, just as they will be standing in the vehicle later on. Now use the 8 mm wood drill, to drill two holes for each box. Drill it directly through both boxes. Use an old piece of wood to prevent the drill holes from tearing out.

12. Securing the boxes in the vehicle

To prevent the fully loaded boxes from sliding when you are stepping on the breaks the boxes should be secured. This can be done in multiple ways, for example with lashing straps. I chose a simple method, which shows great stability. I mounted a **Metal Bracket 105 x 105 mm** to the boxes 1 & 2. These are then connected to the lashing loops in the back of the car (see Illustration 22).

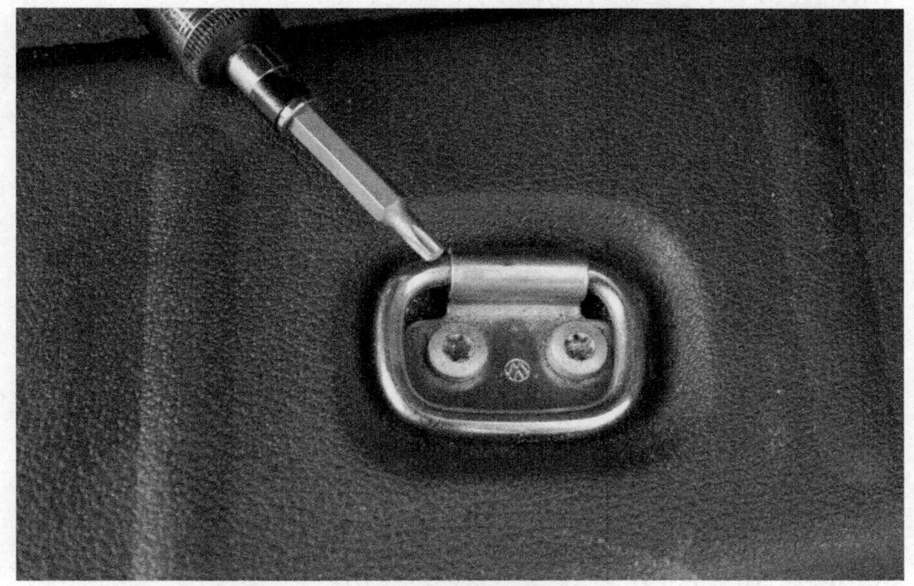

Illustration 22

For this, place the boxes in the vehicle and connect them, just as described in chapter 10. Now push the boxes to the back of the vehicle, so that box 1 & 2 are touching the plastic- cover of the back door (see Illustration 23).

Now unscrew one of the two screws of the lashing loop. Use a Torx bit for this (see Illustration 22). Now use the same screw to fix the metal bracket right at that point (see Illustration 23).

Illustration 23

Now you can mark the drill holes on the boxes 1 & 2, use the 4 mm wood drill and drill the holes where the metal bracket will be mounted.

Use the Hexagon screw 4 x 45 mm each with 1 washer and one wing nut for this.

The boxes are now secured and cannot slide inside the vehicle.

At this point, all the working steps for the boxes are finished!

Congratulations!!!

Just place the two center pieces in the middle and your bed base is done! When reaching your destination, just push the front seats all the way to the front, unfold the folding brackets and assemble the headboards.

13. Mattresses

For a good sleep you need a good mattress. Of course you could choose a inflatable mattress. The advantage for this is that it takes very little storage space. For short trips this is a great idea. But for a more comfortable night and a good sleep a real mattress is crucial.

Therefore I decided to use 3 single mattresses for children. You could also use one large mattress, but using three small ones has some advantages. This way you can stack them and still reach into your boxes when you are driving. When you arrive, you can also use the mattresses as a comfortable place to sit on when you are at the camping space for example.

I chose mattresses which are 140x70 cm. They have a removable cover, which have a zipper and can be washed in the machine. When you remove the cover, it is easy to cut the mattresses with a bread knife into the right shape. To ensure they fit into the car, I had to shorten them a bit (see Illustration 24). I just measured the width of the inside of my car with a tape measure and drew the outlines with a magic marker on my mattresses. Now they fit nicely into my car and they cover the space very well.

I added Velcro fasteners to the mattresses covers, so that they stick together and don't slip apart. I also use a fitted bed sheet to keep everything in place.

I bought the mattresses used on the internet, the total costs were about 60 $.

Cutting template mattresses

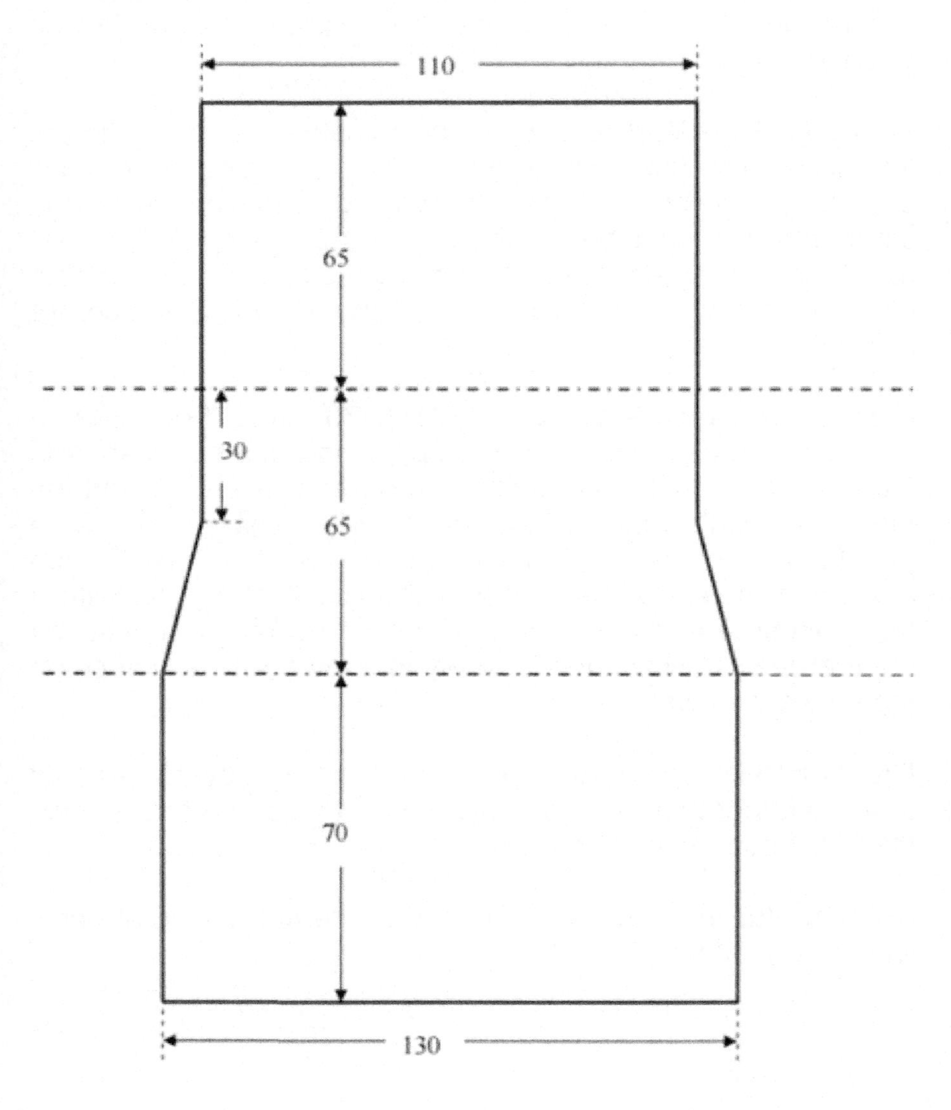

Illustration 24

14. Options

To make my Minicamper more beautiful, practical and cozy, I added some features, which I would like to share with you. It really makes a huge difference while traveling and sleeping in the car and I highly recommend to put some more effort into it. You won't regret it.

Roof liner:
For a cozy atmosphere in my "rolling bed" I took out the roof liner and covered it with a nice and light blanket (see Illustration 25 + 26).

Please be aware that you need a second person to do this! With two people it took us about three hours to this. The blanket is taped to the back of the roof liner with double sided duct tape. Please be aware that when you take out the roof liner, the lamp is still connected with a cable. You can simply disconnect the plug and later put it back on. It took some effort, but the result is really nice and when you are lying in your bed, you won't regret it.

To fix the blanket you can easily use the plastic- screws that hold the roof liner. They can be unscrewed with a Torx Bit. Be careful, these bad boys tend to break easily.

Illustration 25

Illustration 26

Blinds and heat protection:
To keep curious glances outside and to prevent the car from heating up by the sunlight I constructed blinds for the side- and back windows. The idea behind this is quite simple.

Since thermo- blinds are quite expensive, I ordered camping mats that have an isolated side with aluminum foil on it (see Illustration 27). From this mats I cutted out the blinds. I just made some templates out of newspapers and cut them in the right form of the windows inside of the car. I then used the templates on the mats and them out with scissors.

Illustration 27

I then covered the blinds with some nice blankets. It took some patience with the sewing machine, but the end result looks quite ok for me.

The blinds are attached by sucker cups, which I just poked through the cloth.
This is a cheap method, the overall cost were about 30 $. For the windshield I ordered a blind from the internet, which cost about 10 $.

Rear door opener:
To ensure a good oxygen circulation while sleeping, I ordered a door opener (see Illustration 28).
This goes into the rear door lock and the loop for the lock. This way, the door is locked, but a small slot stays open. This is great for hot days.
Costs: ca. 10 $

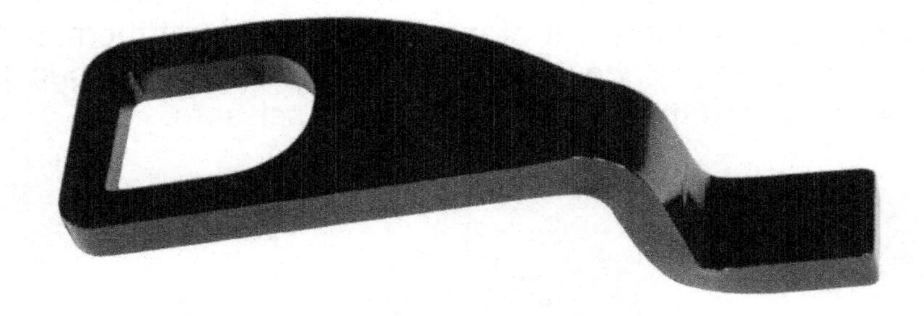

Illustration 28

Lighting:

When all blinds are up, it is really nice and dark inside the vehicle. This allows a good and long sleep, even when the sun comes up. But to be able to read in bed and to have a nice atmosphere I chose an indirect illumination. For this I used a LED- stripe, 100 cm (see Illustration 29). It comes with adhesive tape and it fits perfectly into the backside of the head storage of the caddy, which is above the driver's seat.

The LED strip is powered by USB, for which I use my power bank, which I always have at hand while traveling. This LED strip can be dimmed from really bright to soft lighting and it comes with 10 different colors and a fading effect. So depending on my mood I can change the light from white to blue, to red and so on.

Costs: 10$

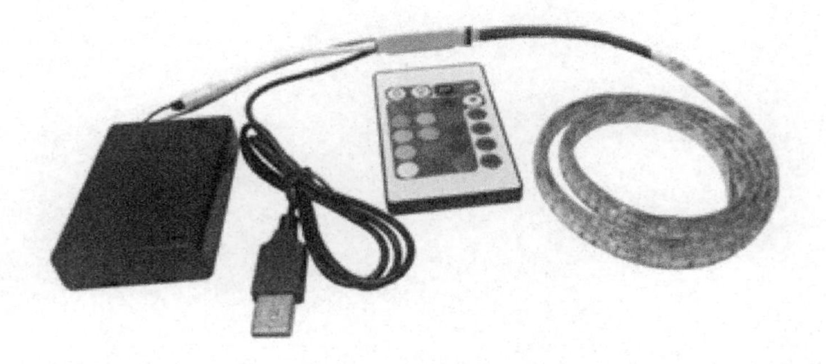

Illustration 29

Tent:

The Minicamper is really ideal for short- and weekend trips. For longer holidays I bought a so called bus tent. This can be attached directly to the vehicle and extends my rolling bed of a good size living room. This allows me to keep the side door of the caddy open even at night, without anyone being able to look inside. Even in heavy rain everything stays dry.

The tent can stand on its own, so you can drive off with your car anytime, for example if you want to go shopping or take a daytrip.

I chose a bus tent from the brand Outwell, the Santa Monica model (see Illustration 30+31).

Costs: about 300 $

Illustration 30

Illustration 31

Camping shower:
In the trunk of my caddy there is a 12 Volt plug, so I bought a 12 volt camping shower (see Illustration 32). The flexible tube is just placed into my water canister. The built- in pump delivers a nice and permanent water jet. The shower head can be hung via a hanger from the back door or it can be fixed via a sucker cup on the car or the window. This will allow you to enjoy a nice shower, even when you are in the wild.

Costs: 10 $

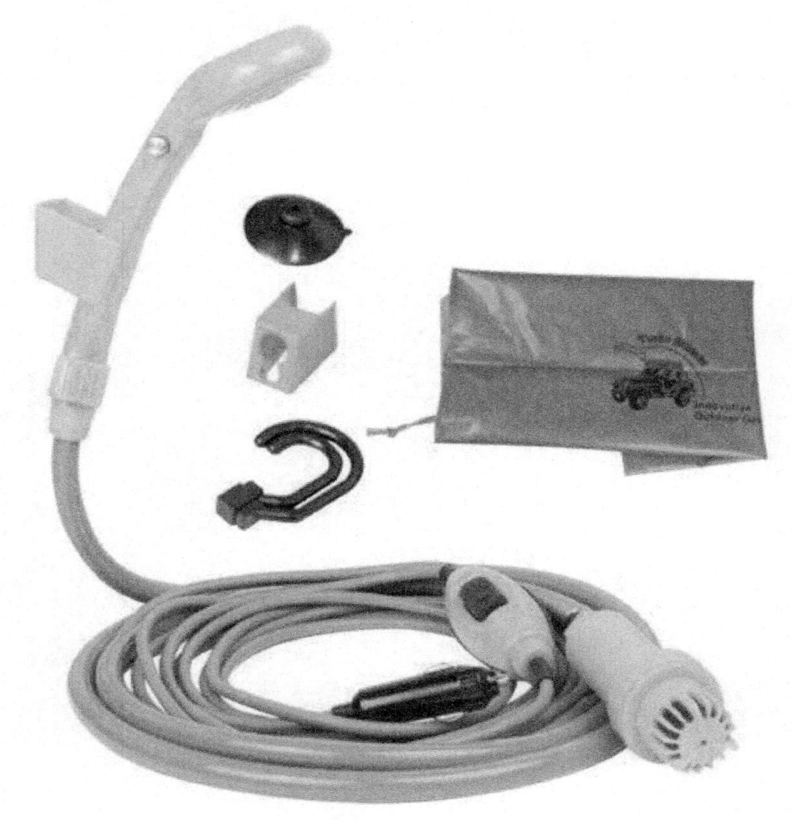

Illustration 32

I hope you enjoyed this book and that it gave you a lot of ideas on how to transform your car into a Minicamper.

If you liked the book, I would be very thankful if you rate this book on Amazon.
This will give me the chance to improve the book constantly for future readers.

It only takes two minutes of your time, but for me it would mean a lot.

Here is the link to your review page:

http://www.amazon.de/feedback

Thank you so much and have fun on the road,

Joerg

Imprint

Joerg Janßen- Golz
Johannes- Huppertz- Str. 36
41352 Korschenbroich
Germany

Tax Identification Number: DE306503308

Used pictures: own material and www.fotolia.de

Printed in Great Britain
by Amazon